FIRST GOD, THEN MAN, NOW AI

Generative Artificial Intelligence

AI & Ajay Malik

Contents

Preface

As the author of this book, I am thrilled to share my passion for the exciting field of Generative AI. I have even taken the innovative step of using Generative AI as a co-author in the creation of this book. The creation story has evolved from divine intervention to human innovation and now to the advancements in Artificial Intelligence. With the advent of generative AI, we are entering a new era where machines not only mimic human intelligence but also surpass it in unprecedented ways. The line between human and machine creativity is blurred.

Now machines can create art and images, write text & literature, write songs and music, and even design molecules. This is due to the advancements in Generative AI, which allows machines to generate new and unique outputs based on a set of inputs and parameters. This technology profoundly impacts various industries such as entertainment, finance, supply chain management, manufacturing, energy, climate change, and transportation.

It is opening up new avenues for artistic expression and creativity. For example, it enables writers to generate new forms of literature, such as poems and short stories, in a fraction of the time it would take to create them manually. In addition, it enables marketing and lead generation teams to target their content precisely. In finance, Generative AI models are used to generate synthetic financial data to train machine learning models and predict market trends. Generative AI is used in supply chain management to optimize logistics and improve inventory management. Generative AI is used in energy to maximize generation and distribution, reduce waste, and improve renewable energy adoption. In climate change, Generative AI predicts and mitigates impacts such as temperature changes, precipitation, and sea level rise. Generative AI is used in transportation to optimize networks, reduce congestion, and improve safety. The benefits of these applications of Generative AI are numerous. They allow for more efficient and effective processes, personalized experiences, reduced costs, improved security, and a more sustainable future. The potential for Generative AI to revolutionize a wide range of fields is immense, and technology will continue to play an increasingly important role in shaping our world.

Let's delve deeper into the inner workings of this cutting-edge technology and learn its incredible potential. From the creative power of GANs and VAEs to the fascinating and sometimes controversial applications of Generative AI, I'm eager to share my insights and experiences with a broad audience.

Through this book, I aim to inspire and educate readers about the possibilities of Generative AI and its impact on our world. Whether you're a tech enthusiast, an AI researcher, or simply curious about the future, I believe this book will provide you with a comprehensive understanding of the field and its exciting future prospects.

Introduction

Generative AI refers to a class of machine learning algorithms that are able to generate new data samples that are similar to or have the same characteristics as a given dataset. These algorithms are trained on a dataset and can then be used to generate new samples that are similar to the ones in the original dataset. Generative AI is often used to create new data samples that can be used for various purposes, such as creating new images, videos, or sounds.

Man is the measure of all things, and with the advent of Generative AI, we are now able to shape and mold the world around us, bringing forth new possibilities and shaping the future in ways that were once thought impossible.

Generative AI has its roots in the early days of artificial intelligence research, with early generative models dating back to the 1950s. However, it has only gained significant traction in recent years, with the advent of deep learning techniques and the availability of large amounts of data. In recent years, Generative AI has made significant strides in the ability to generate realistic images, videos, and sounds. Today, Generative AI is an active and rapidly growing field, with new techniques and applications being developed all the time.

Generative AI can be used to create:

- new data samples that are similar to the ones in a given dataset, which can be used for various purposes such as creating new images, videos, or sounds.

- new data samples that can be used to train other machine learning models, which can help to improve their performance.

- realistic images, videos, and sounds for use in industries such as entertainment, art, and advertising.

- new data samples that can be used in scientific research, such as simulating complex physical systems or creating new molecules for drug discovery.

- new forms of text, literature, poetry, songs, and anything else a creative mind desires

Generative AI is becoming increasingly popular, with the market for Generative AI expected to grow at a CAGR of over 40% from 2020 to 2027.

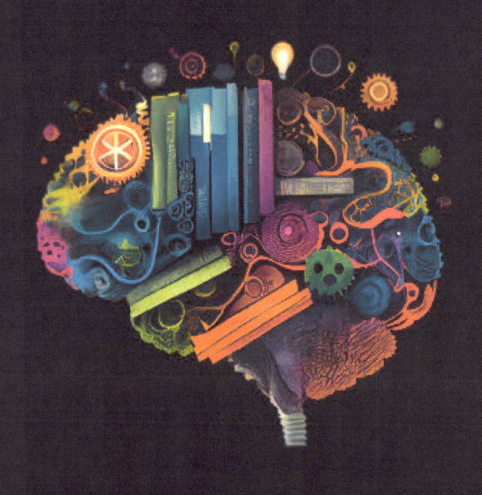

The Cogs and Wheels of Generative AI

There are two main techniques used in Generative AI: Generative Adversarial Networks (GANs) and Variational Autoencoders (VAEs).

Generative Adversarial Networks (GANs) consist of two parts: a generator and a discriminator. The generator generates new data samples and the discriminator determines if the sample is real or fake. The two parts are trained against each other, with the generator trying to generate samples that can fool the discriminator, and the discriminator trying to correctly identify the fake samples. The training process continues until the generator is able to generate samples that are indistinguishable from the real data.

Variational Autoencoders (VAEs) are a type of generative model that consists of an encoder and a decoder. The encoder maps the input data to a lower-dimensional representation (latent space), and the decoder maps the latent representation back to the original data space. The objective of the VAE is to learn a probabilistic mapping between the input data and the latent representation, and to generate new data samples by sampling from the latent space and decoding the samples back to the original data space.

Both GANs and VAEs have proven to be powerful tools for Generative AI and have been applied to a wide range of applications, including image synthesis, audio generation, and anomaly detection.

GANs

Generative Adversarial Networks (GANs) are a class of deep learning algorithms used in generative modeling. They consist of two deep neural networks, a generator and a discriminator, that are trained simultaneously in a zero-sum game-theoretic setting. The generator creates samples and tries to trick the discriminator into believing they are real, while the discriminator tries to distinguish the generated samples from real samples.

The structure of a GAN consists of two main components, the generator and the discriminator. The generator is a neural network that takes a random noise vector as input and maps it to the output data space, such as an image. The discriminator is another neural network that takes an image as input and outputs a scalar value that indicates whether it is a real image or a fake image generated by the generator.

The training process of a GAN starts with randomly initializing the weights of the generator and discriminator. Then, the generator creates a batch of fake images from a random noise vector. The discriminator receives a batch of real and fake images and learns to distinguish between them. After the discriminator is trained, it is used to evaluate the fake images generated by the generator. The generator then updates its weights to produce images that are more likely to be considered real by the discriminator.

This process continues until the generator produces fake images that are indistinguishable from real images and the discriminator is unable to determine whether an image is real or fake. At this point, the generator can be used to generate new, previously unseen samples from the output data space. An example of how a GAN can be used to create a fake video is by training the GAN on real videos and having the generator produce a sequence of images that closely resemble the real video. To do this, the input noise vector can be made to change over time so that the generator produces a different frame of the video at each time step. The generated frames can then be combined to produce a complete fake video.

Here is a high-level overview of the steps involved in using a GAN for creating a fake video:

1. First, a dataset of real videos must be collected and preprocessed to create a training dataset for the GAN. This dataset should consist of a large number of videos that represent the type of video that the GAN will be used to generate.
2. Next, the GAN must be defined and structured. This typically involves defining two separate neural networks: a generator network and a discriminator network.
3. The generator network is responsible for generating new, fake videos. It takes as input a random noise vector and outputs a synthetic video.
4. The discriminator network is responsible for determining whether a given video is real or fake. It takes as input a video and outputs a value between 0 and 1, with 0 indicating that the video is fake and 1 indicating that the video is real.
5. The GAN is trained by alternating between training the generator network and the discriminator network. During training, the generator network is fed a random noise vector and is updated to generate more realistic videos, while the discriminator network is updated to better distinguish real videos from fake videos.
6. Once the GAN is trained, it can be used to generate new,

fake videos by passing random noise vectors through the generator network.

7. To create a fake video using the GAN, a random noise vector is fed into the generator network and the resulting synthetic video is output. This synthetic video can then be post-processed to improve its quality, or it can be used directly as the final fake video.

Here is a high-level pseudocode representation of the training process for the GAN:

```
# initialize the generator and discriminator networks
gen = Generator()
disc = Discriminator()

# specify the loss function for the discriminator
disc_loss = BinaryCrossentropy()
# specify the optimizer for the discriminator
disc_optimizer = Adam(learning_rate=0.0002, beta_1=0.5)

# compile the discriminator
disc.compile(loss=disc_loss,          optimizer=disc_optimizer,
metrics=['accuracy'])
# freeze the weights of the discriminator to prevent updating
during generator training
disc.trainable = False

# specify the loss function for the generator
gen_loss = BinaryCrossentropy(from_logits=True)
```

```python
# specify the optimizer for the generator
gen_optimizer = Adam(learning_rate=0.0002, beta_1=0.5)

# compile the generator
gen.compile(loss=gen_loss, optimizer=gen_optimizer)

# train the GAN
for epoch in range(n_epochs):
  # sample random noise
  noise = np.random.normal(0, 1, (batch_size, noise_dim))

  # generate fake video
  fake_video = gen.predict(noise)

  # sample real video
  real_video = real_video_dataset[np.random.randint(0, real_video_dataset.shape[0], batch_size)]

  # train the discriminator
  disc_real_labels = np.ones(batch_size)
  disc_fake_labels = np.zeros(batch_size)
  disc_real_loss = disc.train_on_batch(real_video, disc_real_labels)
  disc_fake_loss = disc.train_on_batch(fake_video, disc_fake_labels)
  disc_loss = 0.5 * (disc_real_loss + disc_fake_loss)

  # train the generator
  gen_labels = np.ones(batch_size)
  gen_loss = gen.train_on_batch(noise, gen_labels)
```

This is just one example of how a GAN can be used to create a fake video. There are many variations and modifications that can be made to this basic process, depending on the specific requirements of the project.

In conclusion, GANs are a powerful tool for generative modeling and have been applied to various tasks such as image synthesis, text generation, and even video generation. The training process can be challenging, but with the right architecture and training strategy, GANs can produce impressive results that are indistinguishable from real data.

VAEs

Variational Autoencoders (VAEs) are a type of generative model that can be used to generate new data samples from an underlying probabilistic distribution. Unlike GANs, VAEs are designed to learn a continuous, low-dimensional representation of high-dimensional data and generate new samples by sampling from this learned distribution.

VAEs consist of two main components: an encoder and a decoder. The encoder is a deep neural network that takes in high-dimensional data as input and maps it to a lower-dimensional representation, known as a latent code or bottleneck. The decoder is another deep neural network that takes in this latent code as input and maps it back to the original high-dimensional space, reconstructing the input data.

The training process of a VAE involves minimizing the reconstruction loss between the original input data and its reconstructed counterpart. At the same time, the VAE tries to maximize the variance of the latent code to encourage it to capture the underlying structure of the data. This is achieved by adding a regularization term, known as the Kullback-Leibler (KL) divergence, to the loss function.

Once trained, a VAE can be used to generate new samples by sampling from the learned latent distribution and feeding the sample through the decoder network. This process allows the VAE to generate new samples that are similar to the training data, while also allowing for some variation and diversity in the generated samples.

VAEs are commonly used for image synthesis, image interpolation, and representation learning. VAEs are trained to learn a compact representation of the data distribution and then generate new data points from the learned representation.

Here is a code example of a simple VAE implemented in PyTorch for MNIST dataset:

```python
import torch.optim as optim

class VAE(nn.Module):
    def __init__(self, input_size=784, h_size=400, z_size=20):
        super().__init__()

        self.fc1 = nn.Linear(input_size, h_size)
        self.fc21 = nn.Linear(h_size, z_size)
        self.fc22 = nn.Linear(h_size, z_size)
        self.fc3 = nn.Linear(z_size, h_size)
        self.fc4 = nn.Linear(h_size, input_size)

    def encode(self, x):
        h = F.relu(self.fc1(x))
        return self.fc21(h), self.fc22(h)

    def reparameterize(self, mu, logvar):
        std = torch.exp(0.5*logvar)
        eps = torch.randn_like(std)
        return mu + eps*std

    def decode(self, z):
        h = F.relu(self.fc3(z))
        return torch.sigmoid(self.fc4(h))

    def forward(self, x):
        mu, logvar = self.encode(x.view(-1, 784))
        z = self.reparameterize(mu, logvar)
        return self.decode(z), mu, logvar
```

```
model = VAE()
optimizer = optim.Adam(model.parameters(), lr=1e-3)

def loss_fn(recon_x, x, mu, logvar):
    BCE = F.binary_cross_entropy(recon_x, x.view(-1, 784),
reduction='sum')
    KLD = -0.5 * torch.sum(1 + logvar - mu.pow(2) - logvar.
exp())
    return BCE + KLD

def train(epoch):
    model.train()
    train_loss = 0
    for batch_idx, (data, _) in enumerate(train_loader):
        optimizer.zero_grad()
        recon_batch, mu, logvar = model(data)
        loss = loss_fn(recon_batch, data, mu, logvar)
        loss.backward()
        train_loss += loss.item()
        optimizer.step()
    print('Epoch: {} Train Loss: {:.6f}'.format(epoch, train_loss
/ len(train_loader.dataset)))

for epoch in range(1, 11):
    train(epoch)
```

VAEs (Variational Autoencoders) have become increasingly popular in recent years. Unlike GANs, VAEs are trained to learn a probabilistic mapping from an input to a lower-dimensional representation (encoding) and then back to a reconstructed output that approximates the input. The key idea behind VAEs is to use an encoder network to map the input data to a lower-dimensional representation and then use a decoder network to map the lower-dimensional representation back to the original data space.

The training process of a VAE involves minimizing the difference between the input data and the reconstructed output while simultaneously minimizing the difference between the encoding and a prior distribution. The prior distribution is often assumed to be a standard normal distribution, but it can be any distribution that is easy to sample from.

For example, a VAE could be trained on a dataset of face images, learning a low-dimensional representation of the underlying structure of faces. Then, by sampling from the learned latent distribution and passing it through the decoder network, the VAE could generate new, synthetic face images that are similar to the faces in the training data.

Using VAEs to create fake video would involve training a VAE on a dataset of real videos, learning a low-dimensional representation of the underlying structure of the videos. Then, by sampling from the learned latent distribution and passing it through the decoder network, the VAE could generate new, synthetic videos that are similar to the real videos in the training data. Note that this task may require specialized architecture modifications or a larger dataset for better results.

GANs vs VAEs

GANs (Generative Adversarial Networks) and VAEs (Variational Autoencoders) are both generative models that are commonly used in AI. However, there are some key differences between the two, which make them well suited for different use cases.

GANs	**VAEs**
GANs consist of two neural networks, a generator and a discriminator, that compete with each other. The generator tries to produce fake data that the discriminator cannot distinguish from real data, while the discriminator tries to correctly identify whether a sample is real or fake.	VAEs consist of an encoder and a decoder, which are both neural networks. The encoder compresses the input data into a lower dimensional representation, and the decoder attempts to reconstruct the original data from this compressed representation.
GANs are trained using adversarial loss, which is a measure of how well the generator is able to fool the discriminator.	VAEs are trained using reconstruction loss, which is a measure of how well the decoder is able to reconstruct the original data from the compressed representation.

GANs are well suited for generating high-quality, realistic-looking images or other types of data.	VAEs are well suited for generative tasks where the goal is to learn a continuous, structured representation of the data, such as unsupervised representation learning.
GANs are typically more difficult to train than VAEs due to the adversarial nature of their loss function.	VAEs are generally considered to be easier to train than GANs, due to the simpler reconstruction loss function.

In summary, if you are looking to generate high-quality, realistic-looking data, then GANs are a good choice. On the other hand, if you are looking to learn a structured representation of your data, then VAEs are a better choice. In some cases, you may want to use both techniques in combination, for example by using a VAE to first learn a structured representation of your data, and then using a GAN to generate new samples from this representation.

Application	GANs	VAEs
Image synthesis	Good	Good
Image to Image Translation	Good	Good
Image Generation from Text Descriptions	Good	Good
Video Generation	Good	Good
Audio Generation	Good	Good
Anomaly Detection	Not Good	Good
Latent Space Visualization	Not Good	Good
Image Restoration	Not Good	Good

Generating high-resolution images	Good	Not Good
Improving image quality	Good	Not Good
Generating realistic synthetic images	Good	Not Good
Generating synthetic data for anomaly detection	Good	Not Good

In the above table, "Good" means that the technique is well-suited for the application, and "Not Good" means that it is not well-suited.

For Image synthesis and generation, GANs are a good choice because they can generate new, high-quality images that are similar to the training data. VAEs can also be used for this purpose but they tend to produce more blurry images compared to GANs.

For Image to Image Translation, both GANs and VAEs can be used to translate an image from one domain to another. For example, translating a grayscale image to a color image or translating a summer image to winter image.

For Image Generation from Text Descriptions, both GANs and VAEs can be used to generate an image from a textual description. GANs are generally better suited for this task as they can generate high-quality images, whereas VAEs tend to produce blurry images.

For Video Generation, both GANs and VAEs can be used to generate new, high-quality video frames. GANs are generally better suited for this task as they can generate high-quality images, whereas VAEs tend to produce blurry images.

For Audio Generation, both GANs and VAEs can be used to generate new, high-quality audio. GANs are generally better suited for this task as they can generate high-quality audio, whereas VAEs tend to produce less coherent audio.

For Anomaly Detection, VAEs are a good choice because they can be used to identify anomalies in a dataset by reconstructing the data and comparing the reconstruction error with a threshold. GANs are not well-suited for this task as they are designed to generate new data, not identify anomalies.

For Latent Space Visualization, VAEs are a good choice because they can be used to visualize the latent space, which is the underlying structure of the data. GANs are not well-suited for this task because they do not have an explicit encoding of the latent space.

For Image Restoration, VAEs are a good choice because they can be used to restore images by removing noise or artifacts from the original image. GANs are not well-suited for this task as they are designed to generate new data, not restore existing data.

GANs are good for generating high-resolution images, as they can generate new images that are highly detailed and realistic. GANs are also effective at generating synthetic images that are similar to real-world images, making them well-suited for data augmentation. Additionally, GANs can be used to generate synthetic data that can be used to train other models.

VAEs, on the other hand, are not as well-suited for generating high-resolution images. VAEs are better suited for tasks that require reconstructing images, rather than generating new images from scratch. This is because VAEs are trained to reconstruct the inputs they receive, rather than generate new outputs. Additionally, VAEs are typically not as effective at generating realistic synthetic images, as they focus more on the underlying distribution of the data, rather than the appearance of individual images.

Note that this is not an exhaustive list, and the suitability of a particular technique may depend on specific task requirements and limitations.

Current, Potential and Futuristic Apps

3

Generative AI has numerous applications across a variety of fields, including image generation, video generation, and speech generation. Here is a table that outlines current, potential, and futuristic applications of Generative AI:

Applications	Current	Potential	Future
Image Generation	Image super-resolution, style transfer, image synthesis	Image inpainting, semantic image editing, portrait manipulation	3D object rendering, interactive scene generation, virtual reality environments
Video Generation	Video frame synthesis, action classification, video summarization	Video style transfer, video inpainting, video manipulation	Dynamic 3D environments, interactive video games, realistic simulation environments
Speech Generation	Text-to-speech, speech synthesis, accent conversion	Speech enhancement, noise reduction, speech recognition	Voice cloning, speech-to-speech translation, personal virtual assistants

In the current state, Generative AI is being used for image super-resolution, style transfer, and image synthesis, while video generation has applications in video frame synthesis, action classification, and video summarization. In the speech generation domain, text-to-speech, speech synthesis, and accent conversion are some of the most common applications.

Potentially, image generation could have applications in image inpainting, semantic image editing, and portrait manipulation, while video generation could benefit from video style transfer, video inpainting, and video manipulation. In speech generation, speech enhancement, noise reduction, and speech recognition are potential applications of Generative AI.

The futuristic applications of Generative AI are even more exciting, with the potential for 3D object rendering, interactive scene generation, and virtual reality environments in image generation, dynamic 3D environments, interactive video games, and realistic simulation environments in video generation, and voice cloning, speech-to-speech translation, and personal virtual assistants in speech generation.

It is important to note that while these applications are described as current, potential, or futuristic, the actual development and implementation of these technologies can vary depending on various factors such as technological advancements, societal and ethical considerations, and funding and resources.

Case Studies of Generative AI

One example of Generative AI being used in video games is the popular racing game "Forza Horizon 4." In this game, the developers used a combination of deep learning techniques, including GANs, to create realistic images of cars and landscapes. The game features a dynamic weather system and changing seasons, which meant that the developers needed a large number of high-quality images to reflect the different conditions.

To create these images, the developers trained a GAN on thousands of real-life photos of cars and landscapes. The generator component of the GAN was then used to create new, realistic images that matched the style and quality of the original photos. These generated images were then incorporated into the game, where they were used to create the dynamic, constantly changing environments that players experience.

Another example is the use of Generative AI in the music industry. Companies like Amper Music and Jukin Media use Generative AI algorithms to produce new and original music. The algorithms are trained on large datasets of existing music, and then generate new songs based on various inputs, such as style, tempo, and mood. This allows music creators and producers to quickly generate custom music tracks for their projects, saving time and effort compared to traditional music production methods.

A product that exemplifies this application of Generative AI is AIVA (Artificial Intelligence Virtual Artist). AIVA is an AI-powered platform that can compose music in various styles and genres, from classical to jazz to rock. By using Generative AI, AIVA can learn from existing music and then generate original compositions that are in line with a specific genre or mood. This technology has been used by movie and game producers, as well as music composers and songwriters, to create unique and personalized music for their projects.

These are just a few examples of how Generative AI is being used in specific industries and applications, revolutionizing the way we create and experience art, music, and entertainment.

Impact of Generative AI

Generative AI has the potential to revolutionize various industries, including entertainment, art, and advertising.

In the entertainment industry, Generative AI can be used to create new video games with highly realistic graphics, as well as generate new stories, characters, and environments. For instance, AI-generated graphics can be used to create characters that look and act like real people, or to generate landscapes and other environments that are so realistic that players feel as if they are actually present within the game. In the film and television industry, Generative AI can be used to create special effects, such as digitally generated creatures or environments, which can then be incorporated into live-action scenes. This can significantly reduce production time and costs, and increase the quality of the final product. In the art industry, Generative AI has the potential to create new forms of artistic expression. For example, artists can use AI to generate new visual and musical works, or to design and create entirely new forms of digital art. This can provide a new source of inspiration for artists and can lead to new forms of creative expression that would not have been possible without AI.

In the advertising industry, Generative AI can be used to create personalized, targeted ads for individual consumers. For example, Generative AI can be used to create unique, targeted advertisements for each consumer based on their interests, behaviors, and other personal information. This can help advertisers to increase the relevance and impact of their ads, and can lead to higher conversion rates and greater ROI. Additionally, Generative AI can be used to create realistic 3D animations and images, which can be used in advertising to create eye-catching and memorable ads.

Overall, the potential impact of Generative AI on entertainment, art, and advertising is immense. It has the potential to dramatically change the way that these industries create and distribute content, and to open up new forms of creative expression and business opportunities.

Ethics

## 4

The ethics of Generative AI is a complex and rapidly evolving field, as the technology continues to advance and new applications are developed. One of the key concerns with Generative AI is the potential for it to be used for malicious purposes, such as creating fake images, videos, or speech that can be used to spread misinformation or manipulate public opinion. Additionally, there are concerns about privacy and data security, as large amounts of data may be needed to train generative models, and this data could be used to create profiles of individuals or groups that could be used for nefarious purposes.

Another area of concern is the impact that Generative AI may have on employment and the workforce. For example, as Generative AI becomes more advanced, it may become possible to automate many jobs that are currently done by humans, such as writing, design, or content creation. This could lead to significant job loss and economic disruption, and there is a growing need to consider the ethical implications of these changes and to develop policies that can help mitigate the impact on workers and communities.

In general, the ethics of Generative AI is an important and complex area that requires careful consideration and ongoing dialogue. It is important for researchers, policymakers, and industry leaders to work together to ensure that the technology is developed and used in ways that promote the public good, protect privacy and data security, and minimize harm to individuals and society as a whole.

Privacy

The use of Generative AI brings up a number of ethical considerations surrounding data privacy, including:

- **Data Collection**: One of the first and most significant ethical considerations is how data is collected for use in Generative AI models. There is a risk that personal data could be collected without consent, and that this data could be used for malicious purposes, such as identity theft or exploitation.
- **Data Use**: Another ethical concern is how the data collected is used once it is in the hands of Generative AI developers. For example, AI models that generate images or videos could be used to create fake or misleading content, or even to create deepfakes that manipulate individuals' appearances or actions.
- **Data Privacy**: A third ethical concern is the privacy of data used to train Generative AI models. This data may contain sensitive information, such as medical or financial records, and it is important to ensure that it is protected and kept confidential.

- **Data Ownership**: A fourth ethical consideration is who owns the data used to train Generative AI models. There is a risk that personal data could be exploited for commercial purposes, or that it could be shared with third-party organizations without consent.

Consideration	Description
Data Collection	The manner in which data is collected for use in Generative AI models.
Data Use	How the data collected is used once it is in the hands of Generative AI developers.
Data Privacy	Protecting the privacy of data used to train Generative AI models.
Data Ownership	Determining who owns the data used to train Generative AI models.

To mitigate these ethical concerns, it is important for organizations and individuals developing Generative AI models to have clear policies and guidelines in place. This can include regulations around data collection and use, as well as measures to ensure data privacy and confidentiality. Additionally, transparency and openness in the development process can help build trust and confidence in the use of Generative AI.

Bias

Bias in Generative AI refers to systematic and unfair inaccuracies in the algorithms and models developed by AI systems. This bias can be a result of various factors, including the data sets used for training, the algorithms used for generating results, and the decision-making processes of the developers who create these systems.

The impact of bias in Generative AI can be significant, as it can perpetuate and amplify existing societal biases, perpetuate stereotypes and discrimination, and even undermine public trust in the technology.

Examples of Bias in Generative AI

- Gender bias in facial recognition technology: AI systems trained on data sets containing a disproportionate representation of male faces may be less accurate in recognizing female faces and vice versa.

- Racial bias in criminal justice systems: AI systems used in the criminal justice system, such as risk assessment tools, may perpetuate racial biases and result in unequal treatment of different racial groups.

- Economic bias in lending and insurance: AI systems used in financial services may perpetuate economic biases and result in unequal access to credit and insurance services.

Mitigating Bias in Generative AI

To mitigate bias in Generative AI, it is crucial to have diverse and representative data sets, use algorithmic techniques that are specifically designed to reduce bias, and have an ethical and diverse decision-making team in the development process.

- Diverse Data Sets: Using diverse and representative data sets for training AI models can help to reduce the likelihood of biased results. This includes using data from a variety of sources and ensuring that it is representative of the population it is intended to serve.

- Algorithmic Techniques: Algorithmic techniques, such as fairness constraints and adversarial training, can be used to reduce the impact of bias in AI systems.

- Ethical and Diverse Decision-Making Teams: Having an ethical and diverse decision-making team in the development process can help to identify potential biases and address them before they become a problem. This includes involving stakeholders from diverse backgrounds, including individuals from underrepresented groups and experts in ethics and privacy.

In conclusion, ethical considerations surrounding bias in Generative AI are crucial for ensuring that the technology is used fairly and equitably. By taking steps to mitigate bias, we can help to ensure that AI systems are designed to benefit all members of society and are trusted by the public.

Potential –ve Consequences

Here are specific examples to illustrate the potential negative consequences of Generative AI:

- **Fake news and misinformation**: Generative AI models can be used to create fake news stories and manipulate images, videos, and audio recordings. This can lead to the spread of misinformation and undermine trust in news sources. Example: In 2020, researchers at Stanford University used a GAN to generate realistic-looking but fake articles that were shared on social media.

- **Bias and discrimination**: AI models can perpetuate existing biases in the data they are trained on. For example, a model that generates faces might produce images that are predominantly white, male, or young. This can reinforce harmful stereotypes and contribute to discrimination. Example: A study by MIT found that GANs trained on standard facial recognition datasets produced fewer images of dark-skinned individuals, making it more difficult for these individuals to be accurately recognized by the models.

- **Privacy violations**: Generative AI models can be used to generate fake identities and personal information, which can be used for identity theft or to carry out malicious activities. Example: In 2018, researchers generated a synthetic dataset using VAEs that closely resembles real-world financial data. They demonstrated that this synthetic data could be used to train a generative model that can generate realistic but fake financial transactions, potentially enabling fraudulent activities.

- **Intellectual property infringement**: Generative AI models can be used to generate copies of protected works, such as music, art, or literature, which can undermine the value of these works and harm the creators and owners of intellectual property. Example: A GAN was used to generate deepfakes of famous celebrities, which raised concerns about the potential misuse of this technology for malicious purposes such as creating fake news or compromising the privacy of individuals.

- **Economic disruption**: Generative AI models can automate many tasks that were previously performed by humans, potentially leading to job losses and economic disruption. Example: A GAN was used to generate synthetic training data for self-driving cars, enabling the development of autonomous vehicle systems without the need for expensive and time-consuming data collection from real-world environments. This could potentially displace human workers in industries such as transportation and logistics.

These are some of the potential negative consequences of Generative AI, and it is important for researchers, developers, and policymakers to carefully consider these issues and work towards responsible development and deployment of these models.

Best Practices

Here are some key things that you should consider when building or deploying Generative AI:

- **Data Privacy**: When collecting and using data for training Generative AI models, it is important to consider the privacy of individuals. Best practices would include obtaining informed consent from data subjects, using secure methods for data storage and transfer, and following relevant privacy regulations.

- **Bias Mitigation**: Generative AI models can be trained on biased data, which can result in perpetuating and amplifying existing biases in the output of the model. Best practices for mitigating bias in Generative AI models would include using diverse training data, regularly auditing models for bias, and involving a diverse group of stakeholders in the development process.

- **Transparency and Explainability**: Generative AI models can be complex and difficult to understand, which can make it difficult for stakeholders to assess the risks and benefits of using these models. Best practices for increasing transparency and explainability of

Generative AI models would include documenting the development process, including a clear description of the data and algorithms used, and providing information on how the model makes decisions.

- **Safety and Security**: Generative AI models can be used for malicious purposes, such as generating fake or misleading information. Best practices for ensuring the safety and security of Generative AI models would include regularly auditing models for safety and security risks, using secure methods for data storage and transfer, and implementing access controls to limit who can use and modify the models.

- **Collaboration and Stakeholder Engagement**: Generative AI can have a significant impact on individuals and society, and it is important to involve a diverse group of stakeholders in the development and use of these models. Best practices for collaboration and stakeholder engagement would include involving stakeholders in the development process, regularly consulting with stakeholders on the impact of the models, and creating channels for stakeholders to provide feedback on the use of the models.

- **Ongoing Monitoring and Evaluation:** Generative AI models are not static and can evolve over time. It is important to regularly monitor and evaluate the models to ensure that they continue to align with best practices and guidelines for responsible use. Best practices for ongoing monitoring and evaluation would include regularly auditing the models for safety and security risks, assessing the impact of the models on individuals and society, and updating the models as needed to align with best practices and guidelines.

By incorporating these best practices and guidelines into the development and use of Generative AI, organizations can help ensure that these models are used in a responsible and ethical manner that benefits individuals and society.

TL;DR Seven Key Takeaways

1. **Overview**: Generative AI is a subfield of AI that deals with creating new, synthetic data that resembles real data, such as images, videos, and sounds.

2. **Techniques**: Two of the most common techniques used in Generative AI are GANs and VAEs. GANs use two neural networks, a generator and a discriminator, to produce synthetic data, while VAEs use an encoder-decoder architecture to generate synthetic data.

3. **Wide Range Use Cases**: Generative AI has a wide range of applications, including image and video generation, music generation, and data augmentation.

4. **Current, Potential and Futuristic Applications**: Generative AI has already been used in several industries, such as entertainment, art, and advertising, and is expected to be widely used in the future for various purposes such as creating realistic images and videos for video games and movies.

5. **Ethical Considerations**: Generative AI raises several ethical concerns,

such as data privacy and bias, which need to be addressed through responsible use of this technology.

6. **Negative Consequences**: Generative AI also has the potential for negative consequences, such as cybercrime and exploitation, which need to be prevented through responsible use and regulation.

7. **Best Practices and Guidelines**: To ensure responsible use of Generative AI, it is important to develop and follow best practices and guidelines, such as transparency, accountability, and ethical considerations.

Current State and Future Prospects of Generative AI

Currently, Generative AI is a rapidly growing field that is finding applications in a variety of industries such as entertainment, art, advertising, and more. The current state of Generative AI is characterized by the following:

- **Advanced Generative Models:** Currently, Generative AI has advanced to a point where generative models such as Generative Adversarial Networks (GANs) and Variational Autoencoders (VAEs) can produce highly realistic outputs, such as images, videos, and audio.

- **Interdisciplinary Applications**: Generative AI is a highly interdisciplinary field and is finding applications in a wide range of domains, such as fashion, game development, medical imaging, and more.

- **Ethics and Bias Concerns**: While Generative AI has many potential benefits, it also raises a number of ethical and bias concerns that must be addressed.

- **Continued Research and Development**: The field of Generative AI is still in its early stages, and there is much ongoing research and development taking place to further improve the capabilities of generative models.

The future of Generative AI is characterized by the following:

- **Improved Realism**: In the future, Generative AI models are expected to produce outputs that are even more realistic and indistinguishable from real-world data.

- **Increased Adoption**: As the capabilities of Generative AI models continue to improve, it is likely that the use of Generative AI will become even more widespread across a variety of industries.

- **Integration with Other AI Technologies**: Generative AI is expected to play a key role in the integration of other AI technologies, such as computer vision, natural language processing, and robotics.

- **Addressing Ethical and Bias Concerns**: In the future, it is likely that there will be a greater emphasis on addressing ethical and bias concerns associated with Generative AI, in order to ensure its responsible use.

Capability	Current State	Future State
Advanced Generative Models	GANs and VAEs can produce highly realistic outputs	Continued improvement in the realism of outputs produced
Interdisciplinary Applications	Wide range of applications across industries	Increased adoption of Generative AI in industries
Ethics and Bias Concerns	Concerns over ethical and bias issues	Greater emphasis on addressing ethical and bias concerns
Continued Research and Development	Ongoing research and development	Continued improvement in the capabilities of generative models

Improved Realism	Realistic outputs produced by generative models	Further improvement in the realism of outputs produced
Increased Adoption	Widespread use across industries	Further increased adoption of Generative AI
Integration with Other AI Technologies	Key role in integration of other AI technologies	Continued role in the integration of AI technologies

Here's a table outlining current and future state of Generative AI technologies and capabilities:

Technology	**Current State**	**Future State**
Generative Adversarial Networks (GANs)	Used for image and video generation	Advancements in GANs will enable the creation of high-resolution images and videos
Variational Autoencoders (VAEs)	Used for image and text generation	VAEs could be used for real-time 3D object generation
Generative Pre-trained Transformer (GPT)	Used for natural language processing	GPT could be used for generative music and voice generation
Generative models with physical properties	Not yet widely used	Generative models with physical properties could be used for simulations and virtual environments

Generative models for personalization	Limited use in e-commerce and media industries	Increased personalization capabilities could lead to mass customization in industries like fashion and home design
Real-time generation	Currently limited	Advancements in hardware and software could lead to real-time Generative AI
Human-in-the-loop generative models	In development	Human-in-the-loop generative models could enable collaboration between humans and AI in creative processes

The advancements in hardware technology will have a significant impact on the capabilities and potential of Generative AI in the future. With increased processing power, memory, and storage capacity, more complex models can be trained on larger datasets, leading to better results. More specialized GPUs and increased network bandwidth will enable faster training times and more advanced models. Improved energy efficiency will reduce costs and environmental impact, while increased interconnectivity will enable better collaboration and data sharing. All of these improvements will contribute to the continued advancement and growth of Generative AI in the future.

It's important to note that these are potential advancements and the future state of Generative AI is dependent on continued research and development in the field. Nevertheless, the potential applications and benefits of Generative AI are numerous, and it's an exciting time to be involved in the field.

Importance of
Responsible Use

In wrapping up, it's clear that Generative AI has the potential to bring about significant changes to our society and the way we live and work. As we continue to explore the possibilities of this technology, it's important that we remain mindful of its impact and work to ensure that its use is responsible and beneficial for all. With the advancements in AI, we are seeing a new era of artificial intelligence, where machines can now create and generate outputs that were previously thought to be the exclusive domain of human beings. Whether it's generating realistic images, videos, music or even written content, generative AI has the capability to revolutionize our lifestyles and work practices.

However, with the immense potential of this technology, it's important to consider the impact it can have on society, and the need for responsible use. Generative AI is not without its challenges, and we need to ensure that the technology is used ethically, and for the benefit of society as a whole. There are several ethical considerations surrounding generative AI, such as data privacy, bias, and negative consequences, that need to be addressed.

In order to harness the full potential of Generative AI and ensure that it's used responsibly, it's essential that we put in place best practices and guidelines for its use. This can include guidelines for data collection and use, measures to reduce bias and discrimination, and regular reviews and audits of AI systems to ensure that they are being used ethically and with respect for privacy.

The future of Generative AI is exciting, but it's important that we approach it with caution and a commitment to responsible use. As we move forward, it's essential that we continue to engage in open and honest dialogue about the potential impact of this technology, and that we work together to ensure that it's used in a way that benefits society as a whole. The discussion will not be complete without discussing the environmental impact on Generative AI. The energy consumption of AI systems, including GANs and VAEs, is a growing concern in the field of sustainability. Training large AI models, such as GANs and VAEs, can be computationally intensive and consume significant amounts of energy, which contributes to carbon emissions and climate change.

In general, the energy consumption of an AI system is determined by the computational resources required for training, such as the number of processors, the amount of memory, and the duration of training. To reduce the energy consumption and carbon footprint of AI systems, researchers and practitioners are exploring techniques such as model compression, distributed training, and using more energy-efficient hardware.

It is important to consider the environmental impact of AI and work towards developing more sustainable and eco-friendly approaches to AI development and deployment.

In conclusion, Generative AI has the potential to revolutionize the way we live and work, but it's important that we approach it with caution and a commitment to responsible use. By working together to put in place best practices and guidelines, and by ensuring that the technology is used ethically, we can harness the full potential of Generative AI and ensure that it has a positive impact on society.